SALINE DISTRICT LIBRARY
W9-BOF-545

HOW PLANTS GROW

Angela Royston

Heinemann Library
Des Plaines, Illinois

Published by Heinemann Library,
an imprint of Reed Educational & Professional Publishing,
1350 East Touhy Avenue, Suite 240 West
Des Plaines, IL 60018

Designed by AMR Ltd.
Printed and bound in Hong Kong/China by South China Printing Co. Ltd.

03 02 01 00 99
10 9 8 7 6 5 4 3 2 1

Library of Congress Cataloging-in-Publication Data

Royston, Angela.
How plants grow / Angela Royston.
p. cm. – (Heinemann first library) (Plants)
Includes bibliographical references (p.) and index.
Summary: An introduction to plants including descriptions of their Flowers and fruit, spores and cones, roots, stems, leaves, and methods for storing food.
ISBN 1-57572-824-9 (lib.bdg.)
1. Growth (Plants)—Juvenile literature. 2. Plants—Development-–Juvenile literature. 3. Growth (Plants)—Experiments—Juvenile literature. [1. Growth (Plants) 2. Plants—Development. 3. Growth (Plants)—Experiments. 4. Experiments.] I. Title. II. Series. III. Series: Plants (Des Plaines, Ill.)
QK731.R68 1999
571.8'2—dc21 98-45519
CIP
AC

Acknowledgments
The Publishers would like to thank the following for permission to reproduce photographs:
Ardea: p. 5, D. Greenslade p. 22, J. Mason pp. 6, 23, W. Weisser p. 7; Bruce Coleman: A. Potts p. 4; Liz Eddison: p. 26; Garden and Wildlife Matters: pp. 12, 13, 14, 15, 16, 17, 19, 20, 21, 25, 27, K. Gibson pp. 9, 18; Chris Honeywell: pp. 28, 29; NHPA: S. Krasemann pp. 10, 11; Oxford Scientific Films: K. Sandved p. 8; Tony Stone Images: P. D'Angelo p. 24.
Cover photograph: Ken Gibson, Garden and Wildlife Matters
Every effort has been made to contact copyright holders of any material reproduced in this book. Any omissions will be rectified in subsequent printings if notice is given to the Publisher.

Any words appearing in bold, **like this**, are explained in the Glossary.

Contents

Many Kinds of Plants

Plants keep growing until they die. Some plants live for less than a year. But some trees live for hundreds and even thousands of years.

All plants, even the tallest trees, start life as a tiny **seed** or **spore**. **Roots**, **stems,** and leaves all grow from this tiny beginning.

Flowers and Fruit

Many plants use **flowers** to make new **seeds.** **Pollen** from these orange flowers is carried by insects to **ovules** inside other orange flowers.

The pollen joins the ovules to make new seeds. As the orange seeds swell and ripen, they are protected inside a juicy **fruit**.

Spores and Cones

Some plants do not make **seeds** inside **flowers**. Ferns and mosses produce millions of tiny **spores** that are blown by the wind to start a new plant.

Conifer trees have **cones** instead of flowers. The seeds develop inside these woody cones.

A New Plant Begins

Inside each **seed** is a tiny plant that begins to grow when the seed is planted or falls into the **soil.**

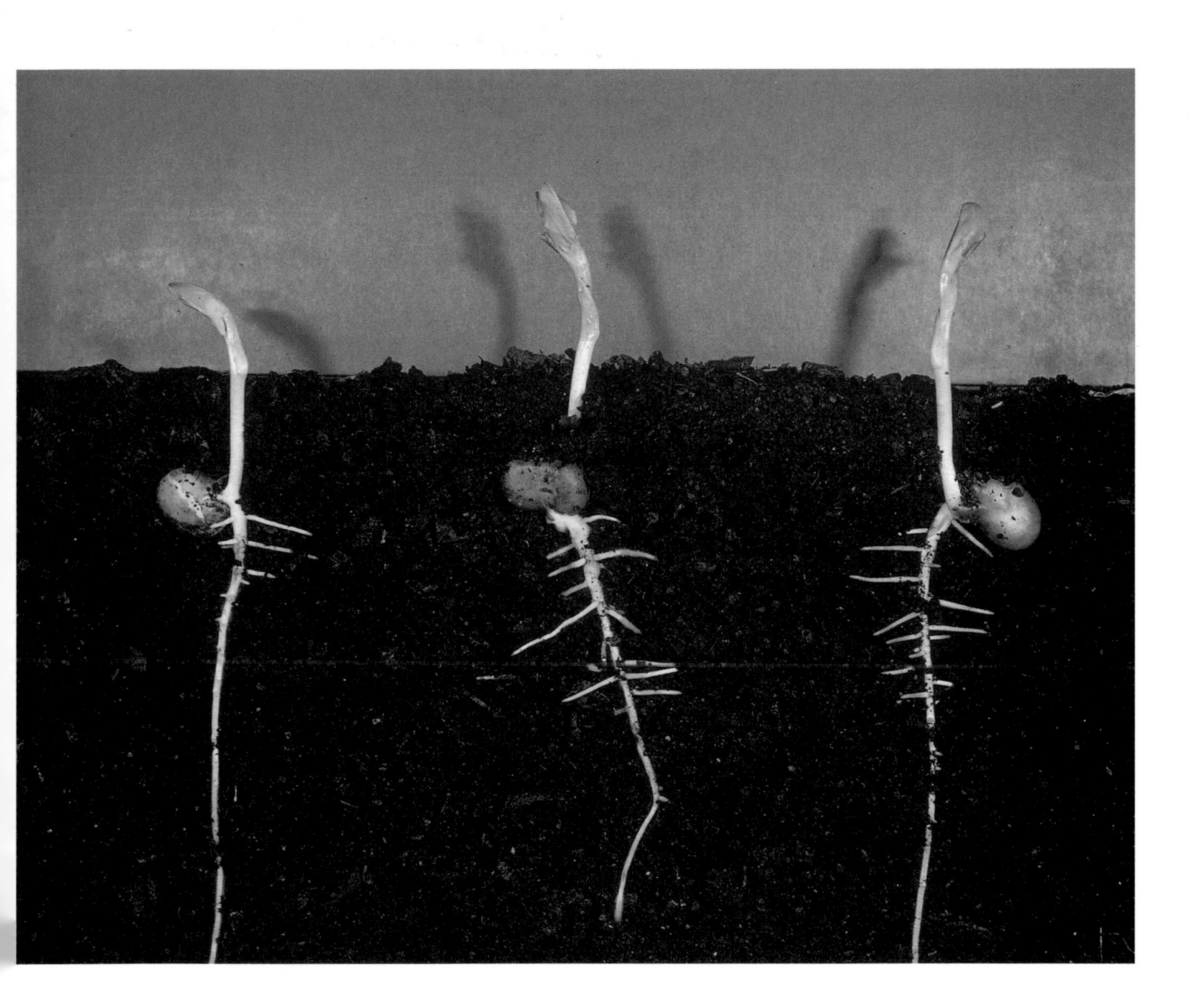

First **roots** begin to grow down. Then a **shoot** pushes up through the soil.

Roots

Green plants need water and sunlight to grow well. **Roots, stems,** and leaves all play a part in keeping the plant alive.

Tiny hairs on the roots take in water and **nutrients** from the soil. Some plants have one big root. Others have a tangled mass of roots.

Stems

The **stem** holds up the leaves and the **flowers**. These sunflowers have straight stems that grow long and tall to lift the leaves up to the light.

Inside a stem there are many tiny tubes that carry water from the **roots** to the leaves. This cactus plant also stores water in its fat stem.

Tree Trunks

Tree trunks are hard, woody **stems**. Trees need strong stems because they grow much bigger and taller than most other plants.

New wood grows every year so the trunk gets thicker and stronger. Bark is hard, dead wood that protects the wood growing underneath.

Climbing Stems

Some plants have long, bendy **stems**. These stems do not support the plant. They climb up something solid, such as a tree or wall, instead.

This climbing plant has curly **tendrils** that twist around a wire or the stem of another plant.

Leaves

Leaves make food for the plant from sunlight, air, and water. The plant turns its leaves toward the sun to take in as much sunlight as possible.

The red tubes in this leaf bring water from the soil. They also take the food made in the leaf to the rest of the plant.

Evergreen Leaves

Some trees and bushes are green all year. Holly has thick, shiny leaves that last a long time.

Conifer trees have small, pointed leaves like needles. These trees lose their leaves a few at a time.

Falling Leaves

Most broad-leafed trees lose all of their leaves in autumn. The leaves may turn yellow, red, or brown before they fall to the ground.

The tree rests during the cold winter weather. It starts to grow again in spring when new leaves grow from the bare branches.

Storing Food

Some plants store food in a **bulb** or swollen **root** in the ground. The leaves die back but in spring, they begin to grow again.

The plant uses the food stored in the bulb until its new leaves begin to make food. This amaryllis needs a lot of food to produce a huge **flower**!

Which Grows Best?

Find out what plants need to make them grow. Put some damp paper towels in the bottom of three soup bowls. Lay some mung beans on top. Keep the paper towels damp and wait for the **seeds** to grow.

Put three bowls near a window. Let one bowl dry out. Add water to the other bowls. Put the third bowl in a box with the lid shut. After a few days, compare the plants. Which have grown best?

Plant Map

A Strawberry Plant
leaf
flower
fruit
stem
roots
An Oak Tree
bark
leaves
trunk
roots

Glossary

bulb	swollen **root** that contains a store of food. Plants that grow from bulbs die back after flowering but grow again the following year.
cone	part of a **conifer tree** that makes new **seeds**
conifer tree	a tree which produces new **seeds** inside **cones**
flower	the part of a plant that makes new **seeds**
fruit	the part of a plant that holds the ripening **seeds**
nutrients	special things a plant needs to grow well
ovule	a female seed or egg cell. An ovule must be joined by a grain of **pollen** to become a fertilized **seed.**
pollen	grains containing male cells that are needed to make new **seeds**
roots	parts of a plant that take in water, usually from the **soil.** Roots also hold the plant up.
seed	contains a tiny plant and a store of food before it begins to grow
shoot	the part of a plant that first grows up out of the **seed**
soil	the ground plants grow in
spore	the cells from which a new fern, moss, or fungus begins to grow
stem	the part of a plant from which the leaves and **flowers** grow
tendrils	thin offshoots of the **stem** of a climbing plant that help to support the plant

Index

More Books to Read

Berger, Melvin. *Plants Set.* New York: Newbridge Educational Publishing. 1995.

Butler. Daphne. *What Happens When Flowers Grow?* Chatham , NJ: Raintree Steck-Vaughn. 1995.

Gibson, Ray. *What Shall I Grow?* Tulsa, OK: E D C Publishing. 1997.

Kalman. Bobbie. *How a Plant Grows.* New York: Crabtree Publishing. 1997.

Maestro, Betsy. *Why Do Leaves Change Color?* New York: HarperCollins Children's Books. 1994.

Rowe, Julian & Molly Perham. *Watch It Grow!* Minneapolis, MN: Children's Press. 1994.